The
Search
For The
Portlaoise
Plane

Joe Rogers

Also by Joe Rogers

STAGE PLAYS
Work of Fiction
The Cross

CHILDREN'S THEATRE
Togher Forest

POETRY
Winter Poets 1992
Caring Poetry Festival 1995
A Collection of Poems 2023

HISTORICAL FICTION
The Diary of a Scullery Maid

AUTOBIOGRAPHY
From an Irish Market Town

COLLECTION OF STORIES
The Rare Auld Times

CHILDREN'S FICTION
The Changeling
The War of the Elves

Dedication

Lovingly dedicated to Kathleen, Requiescat in Pace, my late treasured wife of 63 years who came with me each and every time and helped me enormously in my search for the Portlaoise Plane.

Chapter One

My father, William Rogers, was born in New York city, U.S.A. of Irish parents. His father, also named William, hailed from Skibbereen, Co. Cork, and his mother, Mary Murray, from Ballyfin, Co. Laois. William Rogers, Snr., married Mary Murray on the 27th of January 1884 in the church of St Paul the Apostle on 59th Street and 9th Avenue, Manhattan, New York. Both William and Mary gave their place of residence as New York, and both said they had been born in Ireland. Also listed on their marriage certificate was the names of their parents William's parents recorded as Timothy Rogers and Anne Donovan - her maiden name. And Mary's as Loughlin Murray and Esther Witford - her maiden name. William gave his age as 23 and Mary as 25.

My father was born on the 14th of January 1889 and baptized 3 days later. The first of two children, his brother, Michael, was born on 9th April 1891 and baptized on the 23rd of April 1891. Baptismal records were all important because prior to 1910 an estimated 25% of the population of the New York district were not registered in the civil register of births. Both boys had been baptized in the Church of St. Gabriel, a parish church located at 310 East 37th Street in Murray Hill, (note the name Murray) Manhattan, New York. In 1939 this church was demolished, and the parish closed to make way for the Queens - Mid-town Tunnel linking Manhattan to Queens which was to prove a major problem for my father in later years when due for retirement.

Never having had a birth certificate his baptismal cert proved almost impossible to get until a visiting American priest to Portlaoise learned of his plight, promised to locate the treasured vellum for him and was as good as his word on returning to the States.

In 1893, the two boys, William age 4 and his brother Michael age 2 were brought to Ireland by their mother on the death of their father. As all incoming trans-Atlantic passenger lists had been published, we were able to determine the following:

Name of ship: Germanic
Official number: 71932
Tons as per register: 3149
 Master's name: Edward R McKinstry
Port of Embarkation: New York

Present among the List of Passengers:
Mrs. William Rogers...Mother
Michael Rogers.........Child
William Rogers.........Child

Port at which Landed: Queenstown
Final Destination: Liverpool
Landed October 1893:   Liverpool 15th October 1893

There were no births or deaths on the voyage.

The Germanic had been built in 1874 by Harland & Wolff in Belfast for the White Star Line. Its maiden voyage had been on the 20th of May 1875 from Liverpool to Queenstown and on to New York. The ship would one day hold the record for the fastest Atlantic crossing but on 13th February 1899 she capsized at her New York berth because of too much ice on her decks. Captain Edward Smith, later Captain of the Titanic, was master between 1895 and 1902. In later years when sold and renamed and flying a Turkish flag she was torpedoed and sunk by a British submarine in 1915 during the First World War.

From Queenstown (Cobh, after Irish Independence) Mary and her two children went to her parents' home in Ballyfin. Her husband had died, and she had brought the boys home to grow up with their Grandparents, Loughlin and Esther Murray. William and Michael went to school in Ballyfin, and later William trained as a mechanic and Michael returned to America where he worked for a while with the NYPD and William as mechanic with Frank Aldritt & Sons, Engineers, Maryborough, main agents for Ford and Fordson, Sales and Service. In the early 1900s with the advent of the internal combustion engine enabling the growth of motor vehicle traffic, Frank Aldritt of Maryborough in the Queen's County of Ireland (later Portlaoise in County Laois) established an auto-mobile garage/workshop situated on the corner of Railway Street and Tower Hill in the town of Maryborough.

The business progressed to the point where Frank was in a position to turn his attention to the up-and-coming mode of transportation of the day - Aviation. He gathered a team around him, Louis Aldritt, William Rogers, and John Conroy they set to work to design and manufacture an aircraft, not unlike the Bleriot machine but without manuals or the Internet which was still years in the future their work would certainly not be easy. Long bamboo poles with light wooden spars were used in the wings and would have a canvas cover when completed. The biggest problem with the aircraft was in the design of the engine.

It's necessary to understand that in those early years of the internal combustion engine, both weight and lack of speed were problems. Weight was in the region of 10 to 15 pounds per horsepower output, opposed to only 1 or 2 pounds nowadays and was slow to the degree of only 1000 revs per minute compared with 2/ 5000 today.

These limitations were of no great concern for automobiles that generally travelled at 30/40mph in that era, but proved a considerable problem for aircraft, so much so that Aldritt's first attempt at flight in 1908, in a field that belonged to the parish priest (now CBS sports field) failed. Trials went ahead however, and in 1909 with a redesigned three cylinder in-line water cooled engine of the large bore typical of the times, the plane had a successful flight as reported by the King's County Chronicle on 4th November 1909. The only parts not built by Aldritt's were the crankshaft by R. E.& H. Hall of Salford and a specially designed Bosch magneto bored in Dublin. This turned out to be the first recorded flight in Ireland of a heavier-than-air flying machine, 8 weeks, and a day before the flight of the Harry Ferguson plane on 31st December 1909.

The Portlaoise Plane or the Maryborough Flyer as my father referred to it is a very unique craft because having spent so much of its life either in store or in a museum it has aged well – is the same original aeroplane and not a copy as others of its age will be.

Put in store in Maryborough during WW1 and kept there during the Irish War of Independence, and then the Irish Civil War and WW2 it then disappeared and was missing to the town for over 50 years.

Joe Rogers, a son of William, with a keen interest in aviation, had joined the Irish Army Air Corps where he trained as a Meteorologist and Air Traffic Controller. At Baldonnel (now Casement Aerodrome) Joe learned that a previous Air Corps commander, Colonel James Fitzmaurice DFC was the co-pilot on the Bremen, the first plane to fly the Atlantic from East to West. Colonel Fitzmaurice had grown up in Portlaoise and was a schoolboy there at the time the Portlaoise Plane was manufactured, had witnessed its test flights which inspired him to train as a pilot and go on to win fame as an aviator. I learned while on holiday from England that Portlaoise Plane had gone from the town and the story was that it was in an air museum in England. Well then, I can have a look when I go back. There's an air museum near York and one in Manchester - plenty more up and down the country. Of course, I knew all about the manufacture and history of the Portlaoise Plane or the Maryborough Flyer as it was originally called and I remember as a schoolboy at the Christian Brothers' School, Portlaoise, seeing daily a Propeller over the entrance to the old Aldritt works, corner of Railway Street and Tower Hill. We learned about it as children from Dad and thought what a pity it's not on show up in the Guards' Barrack's yard instead of that old Cannon Gun facing the gates.

Very few people in Portlaoise seem to know about the aeroplane that was made right here in the town - no-one ever talks about it except a few older folk, which is sad because it is part and parcel of Portlaoise history. A very exciting and commendable part of our history and God knows we've not had much of that whereas with almost 800years of resistance within us we've had more than our share of bad history.

TIMELINE OF AVIATION

1485 – 1500   Leonarda da Vinci designed the Aerial Screw (1st ever VTOL machine) & Parachute.

1783   Invention of Hot Air Balloon

1895   Otto Lilienthal uses a biplane

1903 17th of December:
Orville & Wilbur Wright 1st Flight at Kitty Hawk, North Carolina, USA. 12 secs.

1907 17th of August:
Plans for Portlaoise Plane as per
Nationalist & Leinster Times.

1909 25th of July:
Louis. Bleriot First crossing of The English Channel.

1909 4th November:
King's County Chronicle reports flight of Portlaoise Plane.

1909 31st of December:
Harry Ferguson's flight in Co. Antrim.

1919 14th of June:
Alcock & Brown first West to East Atlantic crossing from Newfoundland to Clifden Co. Galway.

1928 12th of April:
Col. James Fitzmaurice first East to West Atlantic Crossing in Bremen with Hermann Koehl & Baron Von Huenefeld.

William Rogers with his daughter Philomena and grandchildren.

Chapter Two

At the end of my five years Regular service with the Irish Defence Force, Army and Air Corps, I transferred to the Air Corps Reserve for nine years and in keeping with Defence Force regulations obtained permission to reside outside the State. I went to Leeds College of Commerce on a British Institute of Management Course and Bradford Technical College to study Woolens, Worsteds and Man-made Fibres and in keeping with my earlier decision I made my way to Manchester's Space and Air Museum contained within the Science And Industry building on Liverpool Road.

Exhibits on show there included helicopters, United States Space Shuttles, British Avro Shackletons and Avro Avians whose main role was to find and sink submarines. There was also an Avro 707A an experimental aircraft manufactured to test flight data which was to be used on the world's first Delta bomber, the Avro Vulcan. The other very impressive plane apart from the Shackleton was the English Electric Lightning capable of 2,500 mph.

The oldest plane there was an Avro Triplane, the first all-British plane to fly in the UK. And the date, wait for it - 1909 - the very same year as our beloved Portlaoise Plane. But sadly, that was the nearest I got to it because there was neither sight nor sound of it in Manchester.

Joe Rogers and his son Kevin visiting
The Manchester Air & Space Museum.

Not that I expected to find it in the very first air museum I went to - I'd never be that lucky. My wife, Kathleen, and one of our sons was with me so we took a few photographs of the different types of aircraft on view and decided that maybe we should have gone to Elvington where there is a large variety of early types of aircraft including replicas of some flying machines. Well that can be our next port of call in a few weeks time.

Following on from Manchester I next visited the Yorkshire Air Museum at Elvington, York, where a large collection of aeroplanes are on show including replicas of very early flying machines but, regrettably, no Portlaoise Plane nor any knowledge of it or where it might possibly be. After York I stayed for a few days with my daughter in Solihull, Birmingham, and while there I went to the Midland Air Museum near the village of Baginton, Warwickshire, next door to Coventry Airport. This is a truly fantastic open-air museum with a very friendly staff and what I call a hands-on museum in that you can climb aboard many of some 50 aircraft, sit in the cockpit of a Vulcan Bomber or visit the Sir Frank Whittle Jet Heritage Centre which is included in the museum. They also had an Avro Anson on display similar to one I had flown in in the Irish Army Air Corps when we photographed part of Ireland's west coast and tested Dublin Airport's Instrument Landing System. However, despite the fact that the Midland Air Museum has so much going for it, there was neither hide nor hair of the Portlaoise Plane.

Vulcan Delta Bomber & instrument panel at
The Midland Air Museum.

My wife Kathleen and I next went to Duxford Air Museum, Cambridge, Britain's largest Air Museum with over 200 aircraft including a variety from different countries. It was here that I was reunited with an old friend from my Army Air Corps days, Spitfire Trainer 161 in its original shade of Green with its Green/Orange Roundel. It was from Duxford that spitfires first flew. Unfortunately once again we came away very disappointed - the Portlaoise Plane was certainly not at Duxford Air Museum.

Joe Rogers with the Spitfire Trainer 161 at Duxford Air Museum.

*Kathleen Rogers & Philomena Reid (William Rogers' daughter)
with Concorde at the Duxford Air Museum.*

Shortly after our visit to Duxford and with our next visit to an Air Museum lined up we received a copy of the Leinster Express from our nephew Pat Rogers. And there on page 21 was the late Frank Aldritt's obituary.

I had known Frank slightly and there in his obituary among details of Frank's life was the story of how his Grandfather had started the Ford dealership garage and built the first aeroplane in Ireland 100 years ago and IT IS NOW IN A MUSEUM IN SUSSEX. Needless to say this piece of information jumped off the page as if Frank himself was anxious for me to go find the plane.

Obviously I was more than keen to follow up on this information and wrote to Mary Kavanagh of the Leinster Express:

Hi Mary,
I hope you are keeping well in these recessionary times.
I read in Frank Aldritt's obituary in the Leinster
Express that the aircraft built in his grandfather's time
is in a Sussex museum. I wonder if you can tell me the
name of the museum as my father, a mechanic, helped to
build it, and I've been searching to try to find it. Quite a
few museums scattered throughout East Sussex and
West Sussex, it would be a big help if you can narrow it
down.
Many thanks,
Joe Rogers.

Back came replies from Mary Kavanagh and reporter Lynda Kiernan with the name of the museum that housed the Portlaoise Plane. Filching Manor Motor Museum. It was a MOTOR museum while we had been searching AIR museums. Lynda Kiernan told me that she had got the information from Frank's daughter, Eavan. I thanked the ladies for their kind assistance and thanks very much also to Eavan.

And so, some weeks later, April 2012 actually, my wife Kathleen and I motored down to Polegate in East Sussex where we met the proprietor, Karl Foulkes Halbard, a charming gentleman and over coffee in the Medieval Manor, Karl explained that his museum exhibits were rather extraordinary in that without exception they were all record breakers, winners or historical firsts of one type or another.

And what a magnificent collection we were then privileged to see and, amazingly, there among Sir Malcolm Campbell's Bluebird K3 in which he took the water-speed record on Lake Maggiore on 1st September 1937; several Campbell cars; Fangio's very own racing car; a selection of other cars including Bugattis, Mercedes and a Formula 1 winner; various record-breaking motor cycles and boats was the exhibit that really set my pulse racing, the Portlaoise Aeroplane on which my father, William Rogers, had worked, the FIRST Plane to be manufactured and flown in Ireland by Aldritts of Portlaoise or Maryborough as it was back then over 100 years ago. What a thrill it was for me to see the massive wings with the two main spars in each wing consisting of rather large round bamboo poles to which a total of 15 cross-members or ribs of a lightweight timber are attached - the canvas covering having long since gone. The fuselage too, minus its outer covering of course, is in remarkably good condition, complete with tail assembly, cockpit, seat, rudder bar and pedals.

Joe Rogers arrives at The Filching Manor Museum.

Karl Foulkes-Halbard with Joe Rogers & The Portlaoise Plane.

Joe Rogers points to a wing of The Portlaoise Plane.

Sir Malcolm Campbell's Bluebird K3 Speedboat at the Filching Manor Motor Museum.

Chapter Three

I wrote post-haste to Laois County Council and to Portlaoise Town Commission as it was then back in 2012, also to Portlaoise Tourist Board telling them the location of the Portlaoise Plane and that its wings, fuselage, etc, were in pretty good shape, would be relatively easy to restore and would they kindly arrange for its repatriation to be promoted and displayed as a major tourist attraction. The history of how Portlaoise had become the birthplace of aviation in Ireland by means of the plane's maiden flight on 4th November 1909 as recorded in the King's County Chronicle. In the interests of tourism and especially that people be made aware, especially Laois people, of the successful and history-making endeavours of their forebears in the county town of Portlaoise. It should all be shouted from the rooftops as it would be in any other country in the world. Days went past, Weeks went past. No reply from anyone. I wrote again pointing out the following:

The King's County Chronicle 4th November 1909 reported:
"It will be of interest to note that during the last week two brothers, Louis and Frank Aldritt, engineers of Maryborough, Queens County, have succeeded in the presence of some friends in covering a short distance in a small aeroplane. The brothers a year ago designed a motor car which is a triumph of mechanical skill. They

possess a high degree of mechanical skill and have refused offers from across the Channel owing to a love of their native town." Please note: OWING TO A LOVE OF THEIR NATIVE TOWN.

Well due to the lack of response so far, I have to wonder does their native town love them enough to bring there plane home and promote it? This was the year 2012 remember.

Weeks passed without any replies from anyone that I'd written to. So I put details of the plane on an Irish website, www. writing. ie/ tell your own story /a treasure uncovered in East Sussex.

In addition I put articles on Twitter and on Facebook hoping to get the authorities interested and bring the plane home.

On 12th July 2012, with still no response from anyone I had the following article published in the Laois Nationalist:

A PLANE THAT TOOK THE TOWN BY STORM DISCOVERED AGAIN.

Back in the early part of the 20th century, Frank Aldritt established a motor workshop in Portlaoise. The workshop was at the junction of Railway Street and

Tower Hill. In 1908 Mr Aldritt designed an aeroplane, making the propeller and engine, helped by mechanic William Rogers and carpenter John Conroy. The first trial was in the Parish Priest's field. (Now the CBS sports field). The first trial was a failure but they persevered with a redesigned engine. The remodelled aircraft was brought to the Great Heath of Maryborough where the plane reportedly flew for several hundred yards before landing again.

Generations of schoolboys attending the CBS on Tower Hill would grow up looking at the frame of the aeroplane which was suspended from the rafters of Aldritt's workshop. Even after the workshop closed down when Aldritts moved to a new modern garage on the Dublin Road, the frame was visible over the closed gates. One of those boys was Joe Rogers, whose father William had worked on the plane. He had grown up on tales told to him by his father about those early days of aviation.

In time the workshop was torn down and replaced, Portlaoise people of that era have often pondered what happened to the aeroplane, amongst them Joe Rogers. You can imagine Joe's excitement then when he walked into a museum in East Sussex and first laid eyes on what was the Aldritt plane.

I kept up my letters but they were all falling on deaf ears unfortunately, so I posted poems re the Portlaoise Plane on Twitter and Facebook and then I noticed that Ireland's Own had published an article in their Easter 2013 Special Edition about what they termed "Ireland's High Flyers" but no mention of the very first plane to fly in Ireland, the Portlaoise Plane.

So once again I put pen to paper and wrote to Ireland's Own as follows:

Dear Sir/Madam,
As a regular reader of Ireland's Own for many years I was rather surprised and disappointed that you made no mention of the Maryborough/Portlaoise Plane in your article Ireland's High Flyers in this year's (2013) Easter Special.

Right enough Harry Ferguson's Northern Irish machine is included but the first aeroplane made and flown in Ireland was that manufactured by Aldritts of Maryborough and flown over the Great Heath of Maryborough. See the Laois Nationalist 10th July 2012 to read a report of the record-breaking machine that put Portlaoise on the map as the birthplace of aviation in Ireland.

Yours Faithfully,
Joe Rogers

Shortly after that I received a message from Fran Aldritt to thank me for my efforts in promoting the Portlaoise Plane, she hoped to visit the museum in Sussex and thanked me for my article in the Laois Nationalist.

I kept up my letters and poems to Facebook, Twitter and again wrote to Ireland's Own in April 2014 when they published my article entitled:

WHY IS MORE NOT MADE OF THE HISTORIC PORTLAOISE PLANE?

Then again in 2014 I wrote to the Laois Heritage Society as follows:

Dear Sir/Madam
Just a short note to compliment you on the Portlaoise Heritage Trail. However, I just wondered why there was no reference to the Portlaoise Plane that put the town on the map as the birthplace of Irish aviation in what is now the Republic of Ireland.

The Society referred to me in their Journal, Volume 7, 2014 but did not say why there was no reference to the Portlaoise Plane on their tourist map of the town.

The Leinster Express in 2016 published my article entitled:

FLYING MACHINE THAT PUT PORTLAOISE ON THE MAP.

I wrote this again obviously to focus some publicity on the plane and in answer to their request for suggestions of how to increase footfall in Portlaoise now that business was drifting away. ***Bring the plane home from the English museum. Shout the achievements of its makers from the rooftops, put it in a museum along with other local attractions, charge a reasonable admission price and erect a Replica of the plane in the Market Square.***

I also learned that Guy Warner was to publish a book about the early days of flying, through publishers Colourpoint for release in 2016. So I contacted him with a few details of the Portlaoise Plane, some of which he already knew, and some I confirmed to him.

Then I received an enquiry from Filching Manor, East Sussex, Karl Foulkes Halbard enquiring about the plane's propeller.

Hello Joe,

Indeed, it was my late father, Paul, who purchased the Aldritt and transported it back to the UK. The plane was housed in the roof of our shed where we lived in Crowborough, East Sussex. In 1987 we moved to

Filching Manor and the Aldritt came with us and is still in our top shed where you saw her on your visit. The article mentions the propeller which to the best of my knowledge did not come with the plane. I wonder if it still exists? I guess my father's first visit to Ireland was around the late sixties when I think he first heard about the plane.

Stay in touch.
Regards
Karl Foulkes Halbard.

I wrote to Fran Aldritt to ask if she had any information re the propeller, and received the following reply.

Hi Joe,
Hope all ok with you,
Only get back to you now, I have had the flu for the past week and only coming right now, thanks for your email I asked my mam about the propeller she doesn't think it is still in Portlaoise, she thinks daddy's brother Donal sold and scrapped a lot of things in the garage the time Aldritts closed (unbeknown to daddy),it's such a pity the plane was not kept in Portlaoise daddy would of been so proud if there was a momento. I don't think many people actually know about it, which is such a shame.

Daddy used to speak so fondly of his uncle Louis and his own father he really loved the time he spent at the garage, at my dad's funeral Willie Farrell and another man from Athy (I can't remember his name) spoke about the plane and remembers seeing it in Aldritts garage, they had spoken about it flying at the Heath.

So unfortunately for the time being the propeller is a mystery for now. I look forward to seeing Guy Warner's book, and I would like to say thank-you for all you have tried to do and the interest in the plane, I can't understand why the local authorities don't have the same interest.

Hopefully talk again soon Joe
And the kindest regards
Fran Aldritt

Guy Warner's book had been released so I bought a copy from the publishers and sent the following message to Fran Aldritt, as we kept occasional contact about the plane:

Hello Fran.
Happy New Year.
Guy Warner's new book, "Pioneers, Showmen & the RFC" was released last month. I bought a copy from the publisher, Colourpoint Books (www.colourpoint.co.uk) and pages 32, 33 and 34 are given over to your family's aeroplane. There is a photo of the inside of the old

workshop of F. Aldritt & Sons, Maryborough and one of
me and the plane in East Sussex. It's another step in the
publication of the plane which hopefully, one day, will
be on show in Portlaoise. I'd love to see a full-sized
replica of it in the Market Square.

All the best for the New Year, Fran.
Joe

And Fran replied:

Hi Joe,
Hope all good with you.
That's great that it has been recognised and been
acknowledged, a lot of it is down to you keeping it alive,
which I'm most grateful for,
Happy new year Joe and all the best for 2017.

On 14 June 2017 I got a message from Stewart Quinn
who told me that he was the producer and researcher for
a film project by name "Waking Portlaoise" which will
shoot in the old Sacred Heart School on Church Avenue
on Saturday 24th and Sunday 25th June. He said he
wanted to contact me regarding the plane my father was
involved in building and to explain where he thought his
work could fit in with the documentary. If I wouldn't
mind giving him a call on such a number.

I replied to Stewart and among other things I told him that the only interest so far in the Portlaoise Plane was from the Irish Air Corps. I'd been invited to a celebration of 80 years of Military Air Traffic Control in Ireland by the Chief Air Traffic Services Officer and while at Casement Aerodrome I'd been introduced to Brigadier General Paul Fry, Commanding Officer of the Irish Air Corps to whom I'd given a copy of my book which has a section devoted to the plane and I also gave him a copy of my newspaper report "The Plane that took the Town by Storm Discovered Again". He was so interested to the extent that he would love to have the Portlaoise Plane as a number one exhibit in the Air Corps Museum.

Stewart thanked me for the information and told me that he'd got a copy of my book from the library, and they would reference all in the film. I replied;

Hi Stewart,
That sounds great. Thanks very much. I'd really love to see a DVD of the film if there is one. I wish you well with it and congratulate you on such a good enterprise. Cheers for now.
Joe

Then as a follow up, on 26 June, I asked:

Hi Stewart,
How did it all pan out? Well, I hope.
You were going to let me know.
Joe

He replied on the 12th of July:

Hi Joe,
Sincere apologies for the delay replying to you and
thank you for your interest in the project.

We are currently logging the rushes for the editor with
the plan to cut a full-length documentary over the next
few weeks/months.

Recording went really well, and we were absolutely
blessed with good weather and the full commitment of
everybody involved.

Michael Parsons made reference on a few occasions to
your book and the "Aldritt/Rogers aeroplane". We now
have to put everything together and fit the pieces of the
puzzle together.

I'll let you know how things develop in the coming
weeks.

Kind Regards
Stew

Perhaps nothing ever came of it because that was the last I heard. Which was a pity really because naturally enough I would have liked to have seen whatever it was they put together about the Portlaoise Plane. My Father was always so happy and proud of his association with the aeroplane and even the newspaper referred to it in my Dad's obituary.

Leinster Express May 1967

Late William (Billy) Rogers

The death recently at the County Hospital of Mr William Rogers, Marian Place, Portlaoise, removed one of the area's most colourful personalities. Billy, as he was fondly known, was a great favourite for many years with guests at Kelly's Hibernian Hotel, Main St., where his genial smile and ever-willing disposition endeared him to all.

Born in New York in the then parish of St, Gabriels, he came to this country at a very early age and led a most colourful career which brought him into contact with all sections of the community in all parts of the country. He sought his birth certificate for many years but owing to the fact that the old parish of St. Gabriels of New York had been incorporated into a larger unit, it was almost impossible to get the treasured vellum. However an American priest on holiday in Portlaoise heard his plight and promised to look up the record which resulted in Billy getting the much-sought after birth cert.

In the early part of the century when motor cars were few and far between, Billy Rogers held what is regarded as the first civil driving licence under the then Irish Free State. Employed by Aldritt Bros he had vivid recollections of early motoring days and was one of those who assisted in the work which led to the building of an aeroplane. By Messrs Aldritts.

During the War of Independence his services as a driver were always needed, and indeed he often told of the many occasions on which he drove Black & Tans, auxiliaries, and the IRA in those stirring times. For a time he was on the postal service delivery and then went into other employment before again taking up duties with the Post Office. He worked with Messrs Jessops, Portlaoise, and in the 50s joined the staff of Kelly's Hotel, where he became so extremely popular.

There was a large and representative attendance at the removal of the remains to SS Peter & Paul's Church, Portlaoise, and at the interment in the new Cemetery. A feature of the funeral was that the coffin was borne by six sons of the deceased, this must surely be unique and is said to be the first occasion ever of its kind in Portlaoise

Rev P Aughney, C.C. Rev P Dunny, C.C. Portlaoise and Rev B O'Connell, Mountmellick, officiated at the funeral.

And from the Laois Nationalist May1967:

One of the town's first motor mechanics and a man associated with the building of the first aeroplane in the country at Portlaoise died on Friday at the Co. Hospital, Portlaoise. Born in New York, Mr William Rogers, Marian Place, Portlaoise, also drove one of the first cars ever in Portlaoise. In later years he was to take charge of the first Mail Car service on the Portlaoise – Stradbally run. Later he was attached to the postal staff at Portlaoise and worked for some time at Kelly's Hotel, Portlaoise.

During the 2WW, known as The Emergency in Ireland, my father was an ambulance driver in the Red Cross and was awarded the Emergency Service Medal by the Minister for Defence.

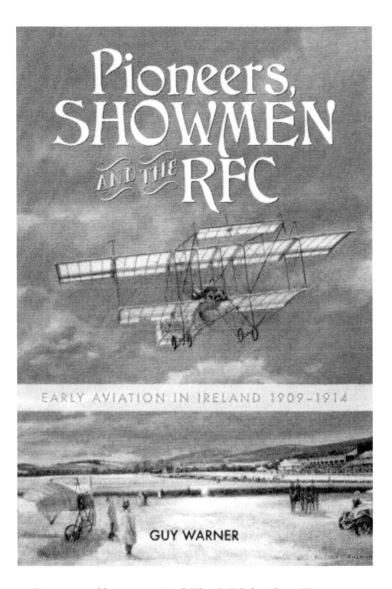

Pioneers, Showmen And The RFC by Guy Warner.

Joe giving a copy of The Nationalist report on the Portlaoise Plane along with a copy of his book to Brigadier General Fry, C/O Irish Air Corps.

Chapter Four

The following is an example of a poem I wrote in an attempt to get the authorities interested enough to bring the plane home and promote it for the good of the town. Always beating the drum, so to speak to try to keep the Portlaoise Plane in the public eye:

THE FAMOUS PORTLAOISE PLANE
By Joe Rogers

Have you heard about the Portlaoise Plane - an Irish Flying Machine,
In 1909 in Laois its first test flights were seen,
Put in store during troubled years, it then from the town was gone -
To be discovered 50 years later by Joe Rogers the Engineer's son.

In Filching Manor Motor Museum in East Sussex by the sea,
Close by Campbell's Bluebird it was there for all to see,
The Pioneer Portlaoise Plane missing for 50 years,
When Joe Rogers found it in 2012 he filled the museum with cheers.

First aeroplane in Ireland - circa 1907 it all began,
With Frank and Louis Aldritt - expert craftsmen to a man,
First flight failed 'cause the engine was too heavy for the frame,
But a new and lighter version brought success to the valiant team.

William Rogers in from Ballyfin was due for early
fame,
When he fixed the problem of propeller spin on the
Portlaoise Aeroplane,
With the Aldritt brothers engineers in charge and
carpenter John Conroy,
They produced the first All Irish Plane to take to the sky
and fly.

The men who brought this dream alive worked long
both day & night,
It was the 'King's County Chronicle' that reported the
plane's first flight,
The Aldritt Brothers and William Rogers assisted by
John Conroy,
Produced the very first aeroplane to take to an Irish sky.

In the North of Ireland in Belfast the Ferguson Plane is
on show,
In the Ulster Folk and Transport Museum for all who
come and go,
And a second Replica of that plane is coming in to land,
Near the A1 Dual Carriageway at Hillsborough all
nicely planned.

So come on Laois County Council make use of the
Market Square,
Tourists will come to Portlaoise to see the aeroplane
there,
A Replica of the Flying Machine made right here in the
town,
Put it in the Square in lights with its makers names all
written down.

So it's fitting that Laois Council have a duty to perform,
An honour to bestow upon these men,
Whose vision and achievement in their workshop long ago,
Produced the Nation's very first aeroplane.

IRELAND

Catharina Day

Cadogan Guides To Ireland.

I was planning to tour Ireland, so I purchased a copy of the Cadogan Guides. And was quite enjoying what I was reading until I turned to page 516. Oh, no! I'm seeing things! It can't be! I read it again and then I started shouting obscenities at Laois County and Portlaoise Town officials for allowing our town to be classified in a tourist guide as HAVING NOTHING TO RECOMMEND IT!

Page 516 glared up at me with the message, and I quote:

The county town is Portlaoise which has nothing much to recommend it except that it is an important rail junction. It has a certain notoriety as there is a well-known prison in the neighbourhood.

Once again, I wrote post-haste to Laois County Council, Portlaoise Tourist Board and to Dublin to the Irish Tourist Board with details of the Cadogan's Guide description of Portlaoise - a sure guarantee of preventing Portlaoise of ever becoming a tourist attraction. How I slighted them for not promoting our number one asset - the Portlaoise Plane...it was their fault entirely that virtually no-one knew of the magnificent engineering feat carried out in Portlaoise that had resulted in the birth of Irish aviation.

I got a reply from the Irish Tourist Board in Dublin, thanking me for the information and they would get in touch with Cadogan. As usual, no answers from Co. Laois.

The following is a copy of my email to Laois Tourism:

Hi,

If you put Cadogan Guides to Ireland into Google or buy a copy of Cadogan Guides you will read what 1,000s of potential tourists are reading in the Cadogan Guides to Ireland and I quote:

"Co. Laois: The county town is Portlaoise, which has nothing much to recommend it, except that it is an important rail junction. It has a certain notoriety as there is a well-known prison in the neighbourhood."
End of quote.

No wonder mainstream tourism is lost to Portlaoise. **Nothing much to recommend it.** How derogatory can you get? What an awful thing to say about a town. I was so annoyed when I read it and wondered why Laois Tourism does not shout aloud from the rooftops that

PORTLAOISE IS THE BIRTHPLACE OF AVIATION IN IRELAND.
THE FIRST AEROPLANE EVER TO BE MANUFACTURED & FLOWN IN PORTLAOISE.

SEE MY REPORT IN THE LAOIS NATIONALIST
1O TH JULY 2012.

The birthplace of aviation in Ireland, surely this fact
should be proudly publicised and special efforts made to
strenuously promote it even to the extent of restoring the
actual record-breaking aircraft which is in a museum in
East Sussex, and a Replica flying machine erected in the
Market Square to become a showpiece for tourists and
towns people. A magnificent reminder of what was
achieved in Portlaoise and would surely prove a source
of inspiration for the future.

One thing is certain, it would negate that dreaded entry
in Cadogan Guides. As a Portlaoise man I sincerely
hope measures will be adopted to right this wrong.

Best Regards
Joe Rogers.

I'd been back to the Irish Air Corps at Casement
Aerodrome by invitation for a celebration of 80 years of
Military Air Traffic Control in Ire-land and because of
my service I was introduced to the Officer Commanding
Irish Air Corps, Brigadier General Paul Fry to whom I
gave a copy of the Laois Nationalist with the story of the
Portlaoise Plane. He was very interested and told me
that he would be proud to have the Portlaoise Plane in
the new Museum that was planned and like others that

I'd spoken to was surprised and happy at the same time
that the plane had survived all the turbulent years in
Ireland and the three known moves it had undergone
since.

I put another poem on Facebook:

THE POOR OLD PORTLAOISE PLANE
By Joe Rogers

A Portlaoise workshop was making an aircraft,
The wings and the cockpit, the wheels and the tail,
The water-cooled engine all made in Ireland,
To be ready to fly a.s.a.p. without fail.

The flight was reported in the King's County Chronicle,
On the island of Ireland, the first plane to fly,
Made in Portlaoise by Frank and Louis Aldritt,
Engineer William Rogers and his pal John Conroy.

It's over a century since that great achievement,
But the Laois & Town Councils just don't seem to care,
With the plane now languishing in a museum in
England,
When it should be on show in the top Market Square.

This was followed by another which I called: Up In The Square.

UP IN THE SQUARE
By Joe Rogers

Have you heard about the workmen in the Portlaoise Market square?
A replica of the Portlaoise Plane is to be erected there,
Just like the airborne sculpture of the Ferguson machine,
But will it be 'landing' or 'taking off ' - let's hope it ain't a dream!

Billy Rogers in from Ballyfin was due for early fame,
When he solved the problem of the propeller spin on the Portlaoise Aeroplane,
With the Aldritt brothers engineers-in-charge assisted by John Conroy,
They produced the first All-Irish Plane to take to the sky and fly.

There's nothing to recommend Portlaoise says the Cadogan Holiday Book!
But the town is full of history for those who take a look!
The Great Heath of Maryborough under a cumulus sky,
Witnessed the flight of the Portlaoise Plane go zoom across the sky.

One hundred years and more have gone and with
interest on the wane,
The Irish Air Corps told me they would love to have the
plane,
When they get their new museum they would love to
install it there,
But made in Portlaoise by men of Portlaoise it should be
in the Market Square.

However, the facts are that in the years since finding the
plane there had only been a total of two what you might
call interested parties - the Irish Air Corps in June 2016
and then on 14th June 2017, a full 12 months after,
Stewart Quinn, Producer of "Waking Portlaoise" had
enquired about the plane for his film. But then on 19th
September 2018 I received a phone call and from the
tone of the gentleman's voice I knew this could well be
what I had waited for - he meant business.

He told me that he and another gentleman were
following up on the Portlaoise Plane and if I'd give him
a ring or my phone number, he would give me an
update. He told me his name was Alan Phelan and he
and Teddy Fennelly would be following up on the
Portlaoise Plane.

I received a phone call from Alan next day when we
discussed the Portlaoise Plane, its location, the condition
it was in, how I'd found it, etc. He asked for certain

information which I said I'd email to him which I did. He told me about an upcoming exhibition launch to celebrate the anniversary of Colonel James Fitzmaurice to which he would send me an invitation.

I sent the email as promised told him that naturally I'd be happy to attend the exhibition launch but as my wife is just home after 3 weeks in hospital and oxygen reliant, possibly for some months, I obviously can't leave her. But I'll be there in spirit and trust you will keep me updated. I told him that I had gone to Filching Manor Museum in April 2012, met Karl Foulkes-Halbard and was delighted to see and photograph the Portlaoise Plane having first learned of it from my father who had helped in its manufacture.

I wrote articles on it for The Nationalist on the 10th of July 2012, an article on an Irish website (www.writing.ie/tell-your-own-story/a-treasure-uncovered-in-east-sussex/) in 2012, and Ireland's Own publication on April 18th 2014.

"The Rare Auld Times" my book published in 2015.Pages 13/14 Portlaoise Plane & pages 267/8 Colonel Fitzmaurice & The Bremen. In June 2016 I gave copies of both the book & The Nationalist to Brigadier Paul Fry, Irish Air Corps, (see photo) and a story of the Portlaoise Plane to The Leinster Express on October the 4th 2016.

I've also put articles re the Portlaoise Plane on t Twitter and Facebook.

The Harry Ferguson aircraft took off on 31st Dec 1909, whereas the Portlaoise Plane was mentioned in the King's County Chronicle on 4th Nov 1909:

It will be of interest to note that during the last week two brothers, Louis and Frank Aldritt, engineers of Maryborough, Queens County, have succeeded in the presence of some friends in covering a short distance in a small aeroplane. The brothers a year ago designed a motor car which is a triumph of mechanical skill. They possess a high degree of mechanical skill and have refused offers from across the Channel owing to a love of their native town.

So, on reflection, Alan, I said, it must be assumed that the Portlaoise Plane was actually airborne 8 weeks before the Ferguson machine, and if as I understand a crash resulted from these early trials, the story as told by the 90-year-old Miss Aldritt, of Ferguson benefitting from it may well be correct. Aldritts eventually developed a lighter engine with which the Heath flight was made.

I got a reply from Alan with an invitation to the October event that year He said he looked forward to the references I had on the Portlaoise Plane and would keep me updated on developments.

I emailed Guy Warner to whom I'd given a few details for his book to ask him if he knew where I could get a copy of the King's County Chronicle backdated to 4 Nov 1909.

Good Morning, Guy.

Hope this finds you in good form. I'm not too bad myself.

I've been reading through your book again - a real gem, full of photographs and so much detail of early flights and flyers. Naturally I have a great interest in Aldritt's 1908 Flying Machine on which my late father, William, worked. On page 33, beneath a photograph of the Aldritt Workshop you quote from an article in the King's County Chronicle, a newspaper familiar to me as my dad had a copy of it which unfortunately has long since disappeared. I would dearly love to have a copy of that newspaper if such is available at all and wonder if you could point me in the right direction? I've tried various Newspaper Archives but without success.
Best for now, Guy
Slan agus Beannacht,
Joe

A View From Above - Donal MacCarron.

Alan had asked me for information on the King's County Chronicle so that's why I'd contacted Guy Warner but in the meantime I replied to Alan as follows:

Hi Alan,

Re The King's County Chronicle. I remember my late father had a copy among all his paraphernalia and souvenirs and I was reminded of it when I came across it in Irish Aviation Pioneers 1909/14.

He also had an early edition by, I think it was Donal Mac-carron, that referred to the flight and in "A View From Above" published in 2000 Aldritt's flying is criticised by local clergy, both denominations, and reference is made to a local lad James Fitzmaurice and how he became enthused by association with all that was happening.

Hope that helps,
Joe

Colonel Fitzmaurice who grew up in Portlaoise, learned to fly in the Royal Flying Corps during WW1, later as Officer Commanding the Irish Air Corps, flew into history as co-pilot on the Bremen, on the first ever East to West crossing of the Atlantic Ocean in 1928. Captain Hermann Kohl of the German Airline Luft Hansa was the other pilot and they were accompanied and assisted on their history-making flight by Baron von Hunefeld.

Adverse winds, snow showers, an oil leak, blind flying in fog at night, but through it all they kept going to finally land safely on the frozen surface of a reservoir on Greenly Island in Canada near the border of Newfoundland and Labrador.

Reaction, worldwide, to the news was both immediate and immense. Decorations were awarded to the heroes, and they were feted for weeks in the New World, celebrations that were renewed and continued when they got back home.

Obviously, the fact that Fitzmaurice was an Irishman gave a tremendous boost to the new Irish State and he was promoted to Major until someone pointed out that the American Charles Lindberg, a Reserve Captain in the US Air Corps, on making the first solo West to East flight the previous year, had been promoted to full Colonel. On learning this, the Irish Government amended the Fitzmaurice promotion to full Colonel also.

The three pioneers flying to Dublin from Germany following their invited tour of Austria, Hungary, and Germany, landed to refuel at Croydon where their reception was anything but friendly or civil. Having to undergo customs examination and asked for passports,

which Fitzmaurice didn't have, he undid his overcoat
and pointed to the British medals on his jacket,
and said, "these are my passport ". Medals from his war
service in the Royal Flying Corps, forerunner to the
Royal Air Force.

Wings Over Ireland – Donal MacCarron.

Chapter Five

I also read in "A View From Above", Mr. Aldritt built a monoplane and took to the air while his wife looked on terror stricken at his antics.

On 28/09/2018, Alan wrote:
A 3rd request, see book mentioned below, Irish Aviation Pioneers, can you find this and reference to the Aldritt aircraft?
Best Regards
Alan

To which I replied:
Yeah, no problem, Alan.   1909 to 1914 Early Aviation in Ireland comes under "Pioneers, Showmen & the RFC" in which Page number 33 is given over to the Portlaoise Plane.

See attached photos.
Fan sláintiuil
Joe

Alan got back to me:
That's great Joe,
I was just about to buy a copy of the book on Amazon. There is a bit more on page 34, can you send me a pic of that too.

Best Regards
Alan.

I replied:
Yeah, Sure, will do.

And then the very next day I thought it was time that I went public again to keep the Portlaoise Plane in the public eye, especially the County Councils.

UP IN THE SQUARE
By Joe Rogers

Have you heard about the workmen in the Portlaoise Market square?
A replica of the Portlaoise Plane is to be erected there,
Just like the airborne sculpture of the Ferguson machine,
But will it be 'landing' or 'taking off' – let's hope it ain't a dream!

Billy Rogers in from Ballyfin was due for early fame,
When he solved the problem of the propeller spin on the Portlaoise Aeroplane,
With the Aldritt brothers engineers-in-charge assisted by John Conroy,
They produced the first Southern Irish Plane to take to the sky and fly.

There's nothing to recommend Portlaoise says the
Cadogan Holiday Book!
But the town is full of history for those who take a look!
The Great Heath of Maryborough under a cumulus sky -
Witnessed the flight of the Portlaoise Plane go zoom
across the sky.

One hundred years and more have gone and with
interest on the wane,
The Irish Air Corps told me they would love to have the
plane,
When their new museum is ready, they'll proudly install
it there -
But made in Portlaoise by men of Portlaoise it should be
in the Market Square.

The Birthplace of Aviation is the town's new claim to
fame -
And the men who made this happen are most worthy of
acclaim:
Frank and Louis Aldritt, William Rogers, and John
Conroy
They made the Plane that then became the first in an
Irish sky.

Then Alan wrote:
Go raibh maith agat a Sheofais
My daughter did a bit of digging at the National Library
and found this.
A copy of the King's County Chronicle with details of
the flight of the Portlaoise Plane.

To which I answered.
Fair play to her, and very well done. I always find
historical papers so interesting. I don't know if you've
read "A View From Above" so I've attached herewith a
few photographs.
Slán leat, Joe.

Then Alan told me that
IT'S COMING HOME.
THE 1900s PORTLAOISE PLANE IS TO BE
RESTORED AND PUT ON DISPLAY IN ITS
HOME TOWN.

Negotiations are ongoing so to avoid price hikes please
treat as confidential. Depending on progress it may be
announced at the following exhibition to which I've
been invited. The story goes that Colonel Fitzmaurice, a
schoolboy at the time of the plane's manufacture,
attended the Aldritt automotive works regularly, just 50
yards from the school.

and was so impressed with the plane's manufacture and flight that he became a pilot and eventually flew into history.

I received an email from a good friend in Portlaoise, Ray Harte, a bit more than a friend, he is married to my niece Geraldine. He sent me a picture from the Dunamase Theatre at the start of an exhibition for the 90th Anniversary of Colonel James Fitzmaurice's crossing of the Atlantic from East to West in 1928. The picture as shown below is of yours truly, taken on the day back in April 2012 in East Sussex when I discovered the Portlaoise Plane in Filching Manor Motor Museum there. I decided to spread the good news to members of my family and sent the following email.

Hi all,
Just to tell you that the Rogers name was in the ascendency in Ireland yesterday when Portlaoise celebrated the 90th anniversary of the first East to West crossing of the Atlantic by air in 1928 by Portlaoise man Colonel James Fitzmaurice. The opening ceremony described how Col Fitzmaurice was inspired by the Portlaoise Rogers Plane on which William Rogers had been the engineer back in 1928 and how his son Joseph had discovered it in a museum in Sussex after it was missing for donkeys years.
All the best for now
Joe

Col. J. Fitzmaurice stamp issued 24th February 1998.

William & Joe Rogers.

The following is the message I got from Ray Harte.

Hi Joe,
You are probably aware of the event and exhibition in
Portlaoise this weekend in relation to 90th Anniversary
of James Fitzmaurice's East/West trans-Atlantic
historical flight.

The launch was on last night and the Portlaoise Plane
got a major mention in the scheme of things. You got a
nice mention in relation to the finding of the plane.
Best Wishes,
Ray.

The opening part of the exhibition went off really well
with a full commitment from everyone. Michael Parsons,
on several occasions, referred to your book and the
Portlaoise/Rogers aeroplane, and they are now looking
forward to the rest of the festival.
Regards,
Ray Harte

To Alan I wrote in answer to his communication.

That's really first-class, Alan. Much obliged.
An excellent presentation, well put together. I'm very
glad everything appears to be going so well. The
photographs are good, and the text tells an exciting
story of the outstanding achievements of Colonel

Fitzmaurice. The flypast looked good. And the Portlaoise Plane also - pictures and press reports and the letter from Aer Lingus quite interesting, confirmed what Karl told me when I went down in 2012. I think it bodes well that there's already been an offer to have it restored.
Well done to you.
Very pleased.
I look forward to it coming home.
Slán agus Beannacht,
Joe.

My email to Filching Manor Museum on the 31st Dec 2018:

Hi Karl

A very happy New Year to you and your family. I hope you are keeping well and have had an enjoyable Christmas. It's hard to imagine that 6 years have passed since I visited your wonderful museum, saw the plane my late father helped manufacture and keep thinking that I must certainly come again. Hopefully some time this coming summer when East Sussex is at its best.

Did you have any success in locating the Aldritt propeller? Old Mrs Aldritt had no idea beyond a faint memory of it on display over the Works Entrance which I too remember as a schoolboy long ago.

All the best for now
Kind Wishes
Joe Rogers

My message to Fran Aldritt 1 January 2019:

Hi Fran

*A very Happy New Year to you and your family. I hope
it's a little more special for you this year now that your
Great Grandfather's plane is back where it belongs in
the town it brought to fame and my promotional efforts
and articles that have kept it in the public domain have
finally proved fruitful. I know that your Great
Grandfather and all who worked on the plane including
my own father would be very happy that the plane will
be displayed and generations to come will know of their
great historical achievement.*

*With every good wish to you and yours, Fran,
Slán agus Beannacht
Joe*

Later that day my friend at the museum, Karl Foulkes
Halbard replied to my query telling me that the propeller
had not surfaced and to the best of his knowledge had
never been at the museum, but the aeroplane had left the
museum and should now be back in Ireland.

I replied to thank him:

Hello Karl
Very good news indeed and thanks for letting me know.
I've written so many articles to promote the plane's
history and been critical of the local council for their
lack of interest and the Tourist Board for their
incompetence - the Cadogan Guide to Ireland stated
Portlaoise has nothing much to recommend it. "What
about the Pioneer Aircraft," I screamed at them, " and
the men of vision who made and flew it?" I knew that
sooner or later they would have to bring it home but
why it has taken so long is beyond understanding.
Congratulations and well done on your success with
Bluebird K3. I saw the photo of the truck with K3 on
board setting out from Filching Manor and I knew you
would do well from my visit to Filching Manor, the
many hours put in and the overall skill involved in your
whole operation up to the first trial sails which I think
were in Surrey.

I hope you do well in the Gordon Bennett, the actual
course is through the towns where I grew up, Kildare,
Carlow, Athy, Stradbally. I remember you have in your
museum the actual time pieces from the original Gordon
Bennett. Of course, we delighted in all your exhibits,
have the photographs and many happy memories and
keenly look forward to another visit even though the
Aldritt has now gone.

Which of course pleases me no end.
Best for now
Joe.

And it would have pleased my father also. The first plane to be manufactured in Ireland. The first plane to fly in Ireland. And he had quite a few firsts himself as we saw in Chapter Three.

Then a message from Alan:

Hi Joe,
Quick update.
We are going to view the Portlaoise Plane at its new temporary location in North County Dublin on Friday. We have representatives from the IAA, Aer Corps, Irish Historic Flight, ex Aer Lingus guys expert at restoring old aircraft and also some Aldritt family. Will take some photos again and report back soon.
Best Regards,
Alan

I thanked Alan for the update. Things were certainly progressing now under his guidance.

I thought back to our first contact in September 2018 and on that occasion, I seemed to sense that at last we might get somewhere. And it was turning out like that. Alan didn't hang about - we were definitely on the move with the Portlaoise Plane near to being restored so that it can be shown to the public.

Then in April 2021 I was invited to take part in a special Zoom event that would feature both the Portlaoise Plane and a talk on Colonel James Fitzmaurice 93rd anniversary of his famous Atlantic flight. So, I sent the following message to my American cousins:

To my Joint Number One Favourite American Cousins: Carol, Jim, Dennis & Anne, Dia is Muire daoibh from Cousin Joe. I just wanted to tell you that there is a special Zoom event today April 12th at 7pm UK time, 2pm New Jersey & Boston time for an hour & a half...it's the 93rd Anniversary of Portlaoise Man Colonel James Fitzmaurice Famous Flight in 1928, the first East to West crossing of the Atlantic Ocean to America and the presentation of the restored Portlaoise Plane to the people of Portlaoise. You are cordially invited to attend, and I regret the lateness of this invitation but to the lateness of my invitation I'm afraid it was outside my control.

If you can manage to join the meeting at some stage you are very welcome. See Join Zoom Meeting link below.

Go dtè tù slàn
Cousin Joe.

And join they did and were of the opinion that Alan Phelan had conducted it all so well their reaction was that it had been a very successful get together and that Alan was someone who could make anyone feel like a someone.

Well, tempus fugit, as they say. Time flies.

In August 2021 I received the following communication:

Calmer conditions hoped for the rescheduled take off of the Portlaoise Plane.

A new date and venue has been set for an event marking the Portlaoise Plane's return to Laois.

The Col James Fitzmaurice Commemoration Committee in association with the Laois Heritage Society is looking forward to welcoming the local community to the showcasing of the Portlaoise Plane, take 2!
The rescheduled event will take place on Sunday, September 12 at 2pm at a new location to minimize weather disruption. The Laois Music Centre (formerly Scoil Mhuire) on Church Avenue, is the new venue. The original event scheduled for Heritage Week was completely booked out, with a long cancellation list. Because of this, the committee asks that if you can no longer make it, we ask that you email

portlaoiseplane@gmail.com so we can free up and reallocate places. If you don't need all allocated tickets, please also let them know.

Anybody attending is asked to arrive 15 minutes before the event begins at 1:45pm. The organisers cannot guarantee entry once the event begins as we must respect our wonderful musicians and other attendees. Just a reminder, due to Covid-19 restrictions, you must check in with one of the stewards at the marked entry and sanitise your hands. Please understand that as everybody must be checked in there may be delays when admitting many people – please be patient, the safety of everyone attending is of the utmost importance to us.

As they are exhibiting in a confined space, people attending are also ask that you kindly adhere to social distancing and as it may not be possible to remain 2 metres apart, we encourage you to wear a mask.

A film crew and photographers will be in attendance to document this event so if you have any issue being in the background, please notify the organisers in advance and let a steward know on the day so we can do our best to accommodate you.

"Many thanks to you all for sharing this special occasion with us – this incredible story continues, and we look forward to having a suitable home for the Portlaoise Plane for public display in due course," said the organisers.

And so it came to pass in September 2021 after an absence of more than 100 years the pioneer Portlaoise Plane made a triumphant return to the town of its birth to a tumultuous welcome from crowds of admirers, well-wishers and aviation enthusiasts who assembled in the grounds of the Laois Music Centre to see and admire the plane that had been just recently restored to its former glory. This indeed was a special day in the Laois Music Centre with specially composed music in honour of Ireland's greatest aviator, Portlaoise's own Colonel James Fitzmaurice.

Councillor Conor Bergin, Cathaoirleach of Laois County Council addressed the crowd: "The Plane was designed and built in Aldritt's Garage Portlaoise by Frank Aldritt and his sons with the help of mechanic William Rogers and master carpenter John Conroy – the first plane to be built and take to the air in what is now the Republic of Ireland –and here today we celebrate the vision, creativity, imagination and originality of those Portlaoise men of the past. But it's also important for us to celebrate the achievement of the people who have rescued this priceless artefact from obscurity.

We must remember Joe Rogers, who first highlighted the continued existence of the plane in a private collection in England; Teddy Fennelly and Alan Phelan who pursued the plane and persuaded the owner to part with it and allow it to return home; Brendan

O'Donoghue and Johnny Molloy who painstakingly worked on the craft to restore it to its shining glory and Tim Costello who informed and inspired all those involved from the start of the project, sharing his knowledge and enthusiasm for all aspects of Irish aviation and in particular in designing the replica engine we see on display here today with the plane."
Michael Parsons, Chairman, Laois Heritage Council, also said a few words: "The Heritage Council encourages national and local exploration and appreciation of Ireland's rich natural, built and cultural heritage. The Portlaoise Plane tells a story of exploration, bravery and derring do, that should make the Aldritt, Conroy and Rogers families very proud of their ancestors.

All of us in Laois and Ireland can join in celebrating this great story of these pioneers of Irish Aviation. The Heritage Council commends Laois County Council for its warm support of the Portlaoise Plane. I personally wish that the plane will soon be on permanent display where it belongs - here in Portlaoise," he said.

It was a great day - a very proud day for me and I applauded the speakers, the musicians, the well-behaved crowds and of course, the restorers but all from afar because much as I would have loved to have been present, it was not possible for me to get to Ireland at that time.

So, I decided I'd celebrate the good news of the plane's presentation to the public with a poem on Facebook.

Well, it might take years to get the plane on show in a museum in Portlaoise whereas a Replica in the Market Square could go up in days where the plane would always be on permanent show.

A BEAUTIFUL FLYING MACHINE IN THE MARKET SQUARE, PORTLAOISE.
By Joe Rogers

It was in the 1980's in the Portlaoise Market Square,
The Laois County Council got rid of old laws there,
Markets that for centuries had been held and drawn a crowd,
Were cancelled by the Council and no more would be allowed.

The people of the town condemned the Council out of hand,
But changed their minds when they realised just what the Council planned,
A search was on for the Portlaoise Plane and if and when it was found,
A replica would go up in the Square so they'd started preparing the ground.

Now at last, they're getting ready in the County Town,
Portlaoise,
And Áras an Chontae's come alive,
The Laois County Council is preparing a Flying
Showpiece,
To honour the men whose plane has just arrived.

The men who made the Portlaoise Plane they were a
special breed,
And they set about their mammoth task determined to
succeed,
No manuals then to guide them and the internet yet
unborn,
Only fault finders to scoff at them and treat their plans
with scorn.

William Rogers in from Ballyfin was due for early
fame,
When he solved the problem of the propeller spin on the
Portlaoise Aeroplane,
With the Aldritt brothers engineers-in-charge assisted by
John Conroy,
They produced the first All-Irish Plane to take to the sky
and fly.

So, it's fitting that Laois Council have a duty to perform,
An honour to bestow upon these men,

Whose vision and achievement in their workshop long
ago,
Produced the Nation's very first aeroplane.

The Kings County Chronicle recorded the flight event,
As it took off into the wind and up in the sky it went,
Frank was in his element and sure Louis just couldn't be
held,
Hurrah for Aldritts' Flying Machine – it goes like a
bird!" he yelled.

There's nothing to recommend Portlaoise says the
Cadogan Holiday Book!
But the town is full of history for those who take a look!
First plane to fly in Ireland and made right here in the
TOWN,
'Twas in a private museum in England 'til Joe Rogers
tracked it down.

The Portlaoise Plane to the town has brought fame,
And it deserves an airborne replica in the Square,
And Frank & Louis Aldritt William Rogers & John
Conroy,
Should have their names emblazoned clearly there,
And the Tourist Board can say with pride, Come and
look at the Market Square,
A replica of the Portlaoise Plane has now been erected
there.

Taking off over St Peter's Church, a beautiful Flying Machine,
It's colours resplendent for all to see, orange and white and green.

So it's fitting that Laois Council have a duty to perform,
An honour to bestow upon these men,
Whose vision and achievement in their workshop long ago,
Produced the Nation's very first aeroplane.

A plane that took the town by storm discovered again

A privilege to see a remarkable piece of Portlaoise history

On the 14th March 2022 I was notified that there was to be a Festival of Flight on 9th April 2022.

Dear All,

The Colonel James Fitzmaurice Commemorative Society is delighted to announce that we are involved in hosting the Festival of Flight 2022 - celebrating the rich history of aviation and exploration in Laois with music events, arts and science workshops, history talks and films. The events take place over the 94th anniversary of the first transatlantic East West flight - a pioneering event that brought honour to a son of Portlaoise Colonel James Fitzmaurice, Ireland's greatest aviator.

The Festival of Flight is organised by the Colonel James Fitzmaurice Commemoration Committee, in partnership with Laois County Council, Laois Heritage Society, Music Generation Laois, the Dunamaise Arts Centre and Midlands Science. The event is funded by Creative Ireland Laois as part of the Creative Ireland Programme 2017-2022 in partnership with Laois County Council and the Heritage Council.

A special word of thanks to you, the reader, for reading My Search for the Portlaoise Plane. To me it's almost as if you had joined with me to locate this extraordinary Flying Machine.

A very special vote of thanks must go to two ladies of the Leinster Express, Mary Kavanagh and Lynda Kiernan for their most welcome assistance in the early days of the search. Also of course to Eavan Aldritt for her invaluable help and to Fran Aldritt for her constant support and encouragement during the years, yes, years, when no interest whatsoever was shown in the Portlaoise Plane by Laois County Council or any of its off-shoots, Tourist Board or otherwise.

Thank you also to Karl Foukes-Halbard, whose family kept the plane safe and in a stable condition for many years, and to Alan Phelan & Teddy Fennelly who finally brought it home.

Praise indeed must go to Brendan O'Donoghue and Johnny Molloy for their excellent and very professional restoration and special mention for Tim Costello who designed and produced a very excellent replica engine.

And so ends The Search for the Portlaoise Plane.

But surely, now as we enter another long wait for the plane to be on permanent show in a museum, a

Life-Sized Replica should be placed in the Market Square where it will constantly be visible to all comers, day and night, throughout the seasons, year after year after year, proclaiming to the world that Portlaoise is the Birthplace of Irish Aviation.

The Portlaoise Plane – Summer 2022.

The Portlaoise Plane – Summer 2022.

*Joe's daughter Kathleen visiting the
restored Portlaoise Plane – Summer 2022.*

Harry Ferguson Sculpture, Hillsborough

*Replica of the Harry Ferguson Plane over the A1 Dual
Carriageway at Hillsborough.*

*It would be great to see a similar statue to The Portlaoise Plane
in Portlaoise town to celebrate the bravery and skill of the men
from Portlaoise who built the first plane to fly in Ireland.*

Printed in Great Britain
by Amazon

19751191R00058